# Follow the Sig[illegible]

by John Serrano

Many signs help us be safe. Here is a traffic light. Red tells us to stop. Green tells us to go.

SOUTH

This sign has four sides. It tells us to look out for deer.

This sign has four sides, too.
It helps us get to the hospital.

EMERGENCY
EMERGENCY

This sign tells us that people are working on the road. It tells us to be careful.

This sign tells us to be careful, too.

This sign has many sides. It tells us to look out for trains.

RAIL ROAD
CROSSING
3
TRACKS

This sign has five sides. It tells us to watch for schoolchildren.

This sign has eight sides. It tells cars to stop. Now the children can go!